The Psychology of Don Quixote

The Psychology of Don Quixote
and the Quixotic Ideal

An Essay by **Santiago Ramón y Cajal**
Winner of the Nobel Prize in Physiology or Medicine

Translated and edited by
Lazaros C. Triarhou, M.D., Ph.D.
Bodossakis Foundation Laureate in Neuroscience

CORPUS CALLOSUM
Thessalonica · Indianapolis

The psychology of Don Quixote and the Quixotic ideal.

A Corpus Callosum book published by

Lazaros C. Triarhou, MD PhD
Professor of Neuroscience
University of Macedonia
Egnatia 156
Thessalonica 54636 (Greece)

Copyright © 2016 by L. C. Triarhou
All rights reserved. No part of this book may be used or reproduced, in whole or in part, including illustrations, in any form by any electronic or mechanical means (beyond that copying permitted by Sections 107 and 108 of the U.S. Copyright Law and excerpt by reviewers for the public press), without permission in writing from the author.

Ramón y Cajal, Santiago, 1852-1934, author
  [Psicología del Quijote y el quijotismo. English]
  The psychology of Don Quixote and the quixotic ideal : an essay by Santiago Ramón y Cajal, winner of the Nobel Prize in Physiology or Medicine / translated and edited by Lazaros C. Triarhou, M.D., Ph.D.
  p. cm.
  Includes bibliographical references.
  ISBN-13: 978-1-5142-4794-5
  ISBN-10: 1-5142-4794-1

  1. Psychology–Philosophy. 2. Art–Philosophy. 3. Education–Philosophy. I. Triarhou, Lazaros Constantinos, editor, translator. II. Title.

Printed in the United States of America

Cover.— *El caballero del ideal y su escudero* or "The knight of the ideal and his squire" (obverse), drawing by Fernarndo Contreras from the magazine *Blanco y Negro*, 1934. Watercolor titled *Cajal el Glorioso* ("the Glorious") by an anonymous artist from a Spanish magazine, 1953 (reverse).

The Corpus Callosum logo is based on a lithograph by the neurobiologist Christofredo Jakob from his *Icones Neurologicae* (1897).

CONTENTS

Foreword · 7

The psychology of Don Quixote and the Quixotic ideal

    Editorial Annotation · 13

    English Translation · 17

Notes · 47
Bibliography · 51

FOREWORD

The Spanish histologist Santiago Ramón y Cajal (1852–1934) shared the 1906 Nobel Prize in Physiology or Medicine with the Italian pathologist Camillo Golgi (1843–1926) in recognition of their work on the structure of the nervous system. Cajal served as Professor at the Universities of Valencia, Barcelona and Madrid and, in 1922, founded the Cajal Institute for Neurobiological Research in Madrid.

At a memorial session before the Society of Neurology and Psychiatry of Buenos Aires, held in November 1934, Professor Christofredo Jakob (1866–1956) assessed the importance of Cajal's neuron theory and its implications for neuropsychiatry and biological philosophy [Jakob, 1935; 1938]:

"Cajal's powerful brain ousted the *ethereal fluid* of the *channeling systems of the brain* and placed us on the stable pedestal of *facts* in lieu of fancies. The clear mind of the great Spaniard was able to sum up anew a century's preparatory work into the grandiose conception of the neuron theory, the quintessence of which rests on the most brilliant discovery by the astute scholar. Cajal was the first to irrefutably demonstrate the free ending of its terminal ramifications. Today this seems trivial, but back then it was the revelation of a new world. Its philosophical importance rests with the elimination of the supposed immaterial fluids and the demonstration of the natural basis of all neuropsychic functions, whence the elaboration of a psychobiology became possible."

The driving force of laboratory experimentation was only one of Cajal's talents [Romero, 1984]. What makes him a genius is the tenacity of many of his scientific and social ideas, which were often ahead of his time [Ramón y Cajal Junquera, 2002]. Furthermore, one notes with delight the consistent references by Cajal not only to the findings of research scientists, but, to the contributions of great philosophers and thinkers [Pasqualini, 1999], proving his erudition, and rightfully earning him the titles of "Cervantes of science" [Cannon, 1949; 1951] and "Don Quixote of the microscope" [Williams, 1954; 1955].

In particular, Williams [1954] argues: "Ramón y Cajal was an outstanding figure in European culture, quite apart from his scientific work. He was also a philosopher, an autobiographer, and above all an incomparable draughtsman and illustrator. An idiosyncratic Spaniard, he was not only a great European biologist; he was a visionary of science, and is rightly called *Don Quixote of the Microscope*."

An acute observer of the literary and scientific scene as well as of the roiling social life of the avant-garde, Cajal emerged in twentieth-century Spain in a further role as philosopher and educator. He even came to stand in some sort as a symbol of national cultural rebirth [Sherrington, 1935].

In his essays, the broader context of *Naturphilosophie*, among several key nineteenth-century neuroscientific concepts, becomes evident [Clarke and Jacyna, 1987]. One detects traits of the Romantic medicine and biology that was current among natural scientists during Cajal's formative years, being at interplay with the convictions of the eventual physicochemical reductionism. In the words of Sir Charles Sherrington, "the simplicity of Cajal's ways and ideas illustrates how singular a mixture he presented of old-time ways and ultramodern science" [Cannon, 1949].

Ramón y Cajal contributed a wealth of writings

beyond neuroanatomy. The *Bibliografía Cajaliana* [López Piñero et al., 2000]—the handy tool for the Cajal scholar—lists 240 books and pamphlets, 271 book chapters and congresses, and 381 articles and notes in journals and periodicals from Cajal's pen, and 718 studies on Cajal written by others, up to the year 2000. Many of his works remain bestsellers in Spain and Latin America, a century later.

Psychology in the work of Cajal received attention in Spanish bibliography, exemplified in the scholarly monograph of Ibarz Serrat [1994]. On the other hand, histology as an expression of Cajal's artistic tendency was elaborated upon by De Felipe [2005].

The picture of Cajal that emerges from his philosophical writings is full and candid. He appears intensely patriotic, but ashamed of his country's administrative inefficiency and scientific backwardness, and determined that Spain should have a place on the scientific and intellectual stage [Taylor, 2008].

The present essay, an evergreen which transcends time, portrays "the catholicity of Cajal's interests" [Courville, 1953]. It is part of a collection of twenty Cajal essays on psychology, art and education that I published recently [Triarhou, 2015]. The text bespeaks the author's inquiry into the human brain and the varieties of its empirical experiences in knowing and behaving, bringing together imagination and passion, and helping us to understand or rethink what we thought we already understood about Cajal's own mind.

Cajal's deceptively light-hearted style occasionally conceals a considerable depth of literary insight [Hoare, 1998]. His language poses the challenges related to the formal parlance, intricate syntax, and regional idioms of *español antiguo*. The translation aims to maintain a balance between fidelity to the original and modern English usage.

I gratefully acknowledge, for their courtesy, the staff at: Instituto Ramón y Cajal, Consejo Superior de Investigaciones Científicas, Madrid; Biblioteca Universidad Complutense de Madrid; Bibliothèque Interuniversitaire de Santé, Paris; Biblioteca Nacional de España; the Library of Congress; and the National Library of Medicine of the United States.

Lazaros C. Triarhou
April 2016

HOJA DE MÉRITOS Y TRABAJOS CIENTÍFICOS

DEL DOCTOR

SANTIAGO RAMÓN Y CAJAL

  e) Coloración de los nidos de Held del núcleo del cuerpo trapezóide, y revelación de sus hilos terminales.
  f) En fin, teñido de numerosas células y fibras del asta de Ammon, *fascia dentata*, corteza cerebral, etc., etc.

N.C. Interpretaciones conjeturales sobre algunos puntos de Histo-fisiología neurológica - Bibl. de "La Ciencia Moderna" - 28-Nov. 1896.- (Ideas expuestas en el prólogo de la anatomía normal de la médula espinal - de Pelaez - P.L.-

  c) Reglas para la educación de la voluntad del investigador.
  d) Exposición de lo que debe saber el investigador de la biología.
  e) Marcha lógica de la investigación.
  f) Consejos relativos á la redacción del trabajo científico, etc.

N.C. Prólogo a la edición de Lluria del discurso dirigido en la Academia de Ciencias.

  i) Exposición de la teoría de la simbiosis nervioso-conectiva consagrada á esclarecer los múltiples fenómenos de la regeneración nerviosa tras las heridas, así como los procesos de crecimiento y modelamiento de los apéndices celulares durante la época embrionaria, etc., etc.

N.C. Psicología de Don Quijote y el Quijotismo - Discurso leído en la sesión conmemorativa de la publ. del Quijote en el Colegio Médico de San Carlos - 9-Mayo 1905.-

Corrigenda (penned by Cajal's hand? His secretary's or assistant's?) to the official 1906 biosketch published under the auspices of the School of Medicine [Ramón y Cajal, 1906]. The three fragments shown (pages 40, 43 and 66 of the booklet, respectively) mark insertions for the *Conjectural Interpretations of Certain Points in Neurological Histophysiology* (1896), the *Preface to the Edition of Lluria of the Speech Delivered at the Academy of Sciences* (1897/1898), and the *Psychology of Don Quixote and the Quixotic Ideal* (1905).

EDITORIAL ANNOTATION

Psicología del Quijote y el quijotismo [1905]

As a visionary in the exploration of the infinitely small, Cajal has been christened "Don Quixote of the microscope" [Cannon, 1949; 1951; Williams, 1954; 1955]. Naturally, the "pen happy knight errant of neurology could hardly have refrained from commenting on the state of contemporary Quixotry, as he saw it and would have wished it" [McMenemey, 1969].

In May 1905, three months after being awarded the Helmholtz Medal by the *Königlich-Preussische Akademie der Wissenschaften* (Royal Prussian Academy of Sciences) in Berlin [Ramón y Cajal Junquera, 2000, pp. 88–92], Cajal spoke at the Medical College of San Carlos "On the psychology of Don Quixote and Quixotry" during the festivities that marked the 300th anniversary of the début of the *Ingenioso hidalgo de La Mancha* [Ramón y Cajal, 1905b; 1905c; 1906].

Thirty-nine years later, publisher José García Perona in Madrid circulated a little hardcover book, 10×15 cm in size and 99 g in weight, bound in a gray-maroon cloth, with a series of opinions on women sincerely granted by the celebrated Don Santiago.[1] Perona had the good sense to complement this *tomito* with Cajal's most interesting—and at the time much ignored—literary work on *Don Quixote and Quixotry*, which the editor, Eduardo Arriaga, felt most worthy of making known to readers, "because it is a mighty knock, a romantic fustigator of the awkward materialism that has been reigning in recent times" [Ramón y Cajal, 1944, p. 89–109].

The following year, *Don Quixote and Quixotry* was incorporated into the *Recopilaciones Cajalianas*, published in the capital city of the Province of Álava, Basque Country [Durán Muñoz and Sánchez Duarte, 1945, p. 53–75], and subsequently in Buenos Aires [Ramón y Cajal, 1954, pp. 50–71]. The essay also forms part of the *Complete Literary Works* published from 1947 onwards [Ramón y Cajal, 1961, p. 1291–310].

Following his habit of not relying on either memory or improvisation, Cajal read his address which was a triumph according to the newspapers of the era. A 14-page separatum of the transcript of Cajal's lecture was printed by Nicolás Moya

[Ramón y Cajal, 1905a; Marañon, 1950, p. 11; López Piñero et al., 2000, p. 32]. The speech exemplified the Cajalian spirit in the best possible manner, and evinced the speaker's deep love for science, as well as his unique vision of Spanish culture [Durán Muñoz and Sánchez Duarte, 1945, pp. 55]. Moreover, it offered a fresh outlook on the hero of Cervantes which was highly appreciated by educators [Escribano, 1924].

Cajal discovered *Don Quixote* at the age of 12 years. In the beginning, he did not seem to enjoy a novel in which the hero suffered so many setbacks. Such an impression dissipated over time such that he later cited Cervantes' immortal opus repeatedly in his literary works. Cajal concurred with the view that Don Quixote was not a madman, but a gentleman with solid ideas who consciously chose to be madly loyal to his convictions and duties. In his speech, Cajal considers the *hidalgo* "an ideal of humanity, magnificence and justice," and suggests that those values, instead of being signs of illness, must always be involved in any true science [Tabarés-Seisdedos and Corral-Márquez, 2001; Palma and Palma, 2012].

A previous English translation of Cajal's essay on Don Quixote, by Craigie and Gibson, was included in their monograph with *Selections from his nonscientific writings* [Craigie and Gibson, 1968, p. 205–22]. An Italian translation was also published [Ramón y Cajal, 2012]. The present English translation adheres, as faithfully as possible, to Cajal's lecture, one more evergreen from the pen of the *maestro*, and incorporates all the original footnotes [Ramón y Cajal, 1905b; 1944, p. 89–109]. To preserve Don Santiago's original tint and fervor, some customary terms were left unaltered, such as *hidalgo manchego* ("Nobleman of La Mancha"), *andante caballero* (a Spanish "Knight Errant"), and *caudillo* ("Commander").[2] Finally, I qualified the latter half of the heading ("Quixotic") with the determinant "ideal"—which is what Cajal construes, and since the attributive "Psychology" should be confined to "Don Quixote."

The Psychology of Don Quixote and the Quixotic Ideal

by *Santiago Ramón y Cajal*

EVERYONE ADMIRES the superb moral figure of the *hidalgo manchego*. Don Alonso Quixano the Good, having converted to an *andante caballero*, by suggestion from ludicrous books of chivalry, represents, as has been mentioned one thousand times, the most perfect *symbol of honor and altruism*. Never has even the Anglo-Saxon genius, which gave so much to imagining energetic and original characters, created a more exquisite personification of individualism and sublime abnegation.

But let me briefly point out the salient psychological characteristics of the protagonist in the immortal novel. As his creator recounts to us, Don Quixote buried himself in his chivalresque novels so avidly that 'he forgot even the care of his estate.' And from little sleep, and much reading and reflecting, his brain became desiccated and his reason became unhinged. Amidst his intellectual and affective exaltation, he arrived at the account, whereby, for the blame of human egoism, the world groans with unrighteousness and dishonor; and thus, passing from idea into action, he abandoned the amenity and comfort of the hearth, and sallied forth on a campaign, resolved to "right the wrongs, correct the abuses, and redress the grievances."

Towards the species, he felt that generous and boundless passion of the great religious founders, and wanted to show it by "exposing himself to peril, by the overcoming of which he would reap eternal fame and renown." He fancied that whatever "notion would increase his honor and serve the Republic" was right and requisite, and did not covet any reward other than the grateful remembrance by posterity, along with the affectionate and pious glance from the *Señora* of his dreams. When in his dolorous misfortunes he felt defeated by ominous destiny, he did not sense pain in the body, but in the ideal. However, the defeats did not quench his faith; he thought that he was persecuted by envious and malign enchanters, and hoped to wear them out by the forces of perseverance and heroism or by the help from propitious and generous genies to his ill-fated valor. In vain, the poised and sensible Carrascos and Mirandas, guardians of the laws of common sense, cautioned him about danger and called him back to the prosaic and bitter reality; Don Quixote did not listen to them, and if at times he consented to engaging in discussions with them, he was solely yielding to the inevitable rules of courtesy and good upbringing. What could they say which exceeded the sublime ideal that he was carrying in his brain? Compared to the grandiose and mirific dream, where men were heroes of legend, Nature a golden lattice woven by fairies, women archetypes of elegance and sovereign equipoise, what was the pallid and miserable real world worth? An inner, intense, restricted and recluse life absorbed him; the secluded and engrossed life of the larva, occupied, amidst the roar of the thunder and the fury of the wind, in impassively spinning the golden cocoon of glory...

All great dreamers aspire to realizing their dreams and to vesting their chimeras in flesh and blood by launching on the world a human type different from and superior to the current type, the creator of a powerful current of life that will sweep the barriers raised by sentiment, interest, and tradition. We could say that this is the idea which aspires to curdle in essence, and which, surging in the brain like a distant echo of reality, strives to hark back to its spring, and to rise as a tyrant and master of Nature herself.

This important psychological law, well understood by Cervantes, becomes fulfilled in Don Quixote. He further cherishes a luminous fantasy and longs to live it, and to make others experience it, embellishing and ennobling the earth with its magical sparkles. During his ardent apostolate, he will not resort to suggestion and miracle (those dialectical resources of the meek religious propagandist), but, instead, to the forces of contradiction and the rigors of the blade. No cowardly compromises with the insidiousness and iniquities of the powerful. Consciences are made of hard rock and they should be sculpted with thrusts from the lance.

And he will sculpt them in accordance with *the ideal model of honor that he picked from the stories of heroism*. Because Don Quixote, in addition to possessing a hypertrophic ego, brimming over with will and dynamism, feels fortified by that blind faith in luck, a faith characteristic of the great conquerors of souls and lands.

Had Cervantes not added certain pathological characteristics to such an admirable incarnation of religion, duty and altruism, then Don Quixote's type, despite being of Cyclopean contexture, would have been reduced to the modest proportions of a practical philosopher, somewhat exalted and imbued with arrogant confidence in his lucky star and the supremacy of his mission. But Cervantes (we must not forget) suggested, above all, a work of literary polemic. Wanting to fence against the books of knighthood with the potent sabre of ridicule, he deemed it indispensable, to that end, to somewhat tarnish and curtail the amiable figure of the ingenious hidalgo with the stigma of insanity, his astute and spirited *noûs* becoming prey to and the plaything of illusions, hallucinations, obsessions, and delirious ideas.

More than once I have wondered: "Why did Cervantes not make his hero sane?" The spirited and eloquent defense of realism in the sphere of art does not necessarily require the insanity of the caballero of the ideal. I concur, nonetheless, that a merely philanthropic, even though passionate and vehement, Quixote would not have gladly abandoned the *dolce vita* and the goodies of bourgeois life in order to sally forth on risky and daring adventures. Even in the case that the eagerness for glory and the longing for justice were strong enough to detach him from his shack, leading him to the strenuous soldier against egoism and treachery of the world, would his feats have given rise to such material of artistic labor so as to forge the epic, marvelous, and astonishing episodes that we all admire in the immortal book, and which speak so highly of the ingenious sovereign and creative vein of the prince of our prose writers?

Without a doubt, owing to this compulsory mental derangement of Don Quixote, which impelled him to provoke the most bizarre and dangerous episodes, the general tone of the novel is profound melancholy and disconsolate pessimism. In vain does the reader, moved, seek serenity, becoming aware that Cervantes did not personify in the *Caballero of the Rueful Countenance* but delirious, inconsistent, and implausible chivalrous compositions. Carried away, regretfully, by the generalizing trend of reason, we are assaulted by the fear that the anathema, which hangs over the romantic art in the mentioned work, is extended to domains alien to the plan of the sovereign artist. And we wonder, with restlessness in our soul and tears in our eyes: "What? Would all the lofty idealisms of science, philosophy and politics also be condemned to perish irremissibly? Is it only reserved for dementia to confront the grand heroisms and the great humanitarian endeavors?"

And such a melancholic and bleak emotion reaches acuity when we see how, at the hour of death, the sublime insane, converted by this time into Alonso Quixano the Good, brusquely recovers logic to proclaim the sad and enervating doctrine of resignation before the iniquities of the world. "In last year's nests there are no birds this year," he tells us with a frail voice, in which there appear to vibrate the rattles and pangs of death. Bursting with infinite disillusion, which announces to us how the paradise of peace and fortune and the dreamy golden age that Humanity yearns for the present or for the not too distant future, represents a most remote past that will not return!…

It would be foolish not to acknowledge that, barring the general, righteously passionate tone, a sane and legitimate humoresque abounds and romps in the Cervantean epos. What other cause does the picturesque and delightful type of Sancho represent, but the emotional artistic counterpoise of the querulous and wretched *Caballero of the Rueful Countenance*?

In a faithful reflection of life, pleasure and dolor, those two antipodal and alternating emotions, succeed each other in the immortal novel, just as they do in the cinematograph of human consciousness. Nonetheless, like the fruit that has a sweet rind and a bitter kernel, the acidity in the Cervantean creation is deep, and the sweetness superficial. Certainly, there are adventures and conversations of an incomparable comic visage; but, in spite of the pious intention of the author, beneath the ingenuous and white mask of the *pagliaccio* run silent tears, like the stealthy creek that glides beneath the sunlit snow.

How was there forged, in the fiery imagination of Cide Hamete, such a felicitous and artistic contrast? In virtue of what psychophysiological conditions did such a serene, Quixotic and optimistic writer place into his opus that tinge of sorrow and bitter pessimism? Arduous and rather hard questions, for the solution of which it would be indispensable to understand all the gyri and sulci of the complicated mind of Don Miguel, besides the emotional shocks, episodes and incidents that shook and indoctrinated him during the dreary years that preceded the brilliant conception.

In all this, there is no lack of valuable matter which allows, if not to solve the problem, to at least formulate a certain more or less plausible possibility. Such data—gathered by the penetrating analyses of our foremost critic Menéndez y Pelayo, by the diligence and wisdom of Revilla y Valera, by the recent piece, so copious, artistic and evocative, of Navarro Ledesma, and by the felicitous observations of Unamuno, Salillas, and numerous other expert and devoted Cervantists—teach us that Cervantes, save the realistic parenthesis during which he planned and wrote the immortal book, was always an incorrigible Quixote in action and a romantic poet in his emotions and thoughts.

What occurred, then, so that the amputee of Naupactus would abandon the cult of his artistic ideals? This is easy to deduce, and, moreover, it appears to be confirmed by not a few critical studies.

Cervantes was born and reared with lofty and noble ambitions. A hero at Naupactus, he dreamt of the glory of the great *caudillos*; a sentimental and amatory writer, he longed to bear the poet's crown; an upright and diligent functionary, he aspired perhaps to financial prosperity or, at least, the *aurea mediocritas*; in love in Esquivias, he thought to convert his life into a sweet and perdurable idyll. But, alas, the implacable Destiny transformed his illusions into disillusions, and, just around the peak of life, he lived forgotten, solitary, indigent, captive and disgraced...

Great disenchantments demagnetize the best focused wills and deform even the most incorruptible characters. This is what happened to Cervantes. From that dismal chaos of the Sevillian prison, where one would be drawn to the end of chiseling on the genius as many wounds, anguishes and miseries as can torment and degrade a human creature, surged a new book and a renewed man, the only one capable of writing such a book. A work with no par, amassed with tears and the flesh of genius, where it entirely deplenished an afflicted soul, disenchanted with being alive!

Its pages offer us the synthesis of life, that is to say, lights and shadows, summits and abysses. Like in the glade of a forest, we see down below the blackness of the vegetal mulch, formed by the debris of illusions and the residues of hope (the proper food of literary genius); on top of the earth, erect and staring at the sky, the robust trunks of exalted ideas, noble aims and sublime aspirations; and above, bathed in the blue atmosphere, the foliage of natural language, pure and colorful, the delicate flower of poetry, and the acrid fruit of experience.

Many have remarked that the supreme Cervantean creation is the most perfect, the ultimate, the insuperable book of chivalry. But in such a judgment, paradoxical at first sight, and clashing with the confessed objective of the work, as well as the explicit declarations of Cervantes himself, I can only see the tacit affirmation whereby the figure of the protagonist is so sovereign, so lovingly felt and sketched, that perforce the author should himself have borne some and even much of Quixote. Such perfect and vivid human portraits do not rise from the brush, unless the painter has repeatedly looked himself in the mirror and focused on the hideouts of his own consciousness. But, after acknowledging such a spiritual kinship between Don Quixote and his author, it is mandatory to also concur that, in the incomparable novel, in the turns of a certain *ritornello* and in the ancient chivalrous adventures, are camping, and become externalized with eloquent accents, the despondency of the pathos of the ideal, the dolorous abandonment of a tenaciously cherished delusion, the *mea culpa*, somewhat ironic perhaps, of disenchanted and defeated altruism.

To keep his mind serene and his imagination lively and plastic, it is necessary for the unfortunate poet to evoke, from time to time, smiling images, capable of hiding and adorning the dismal fundus of consciousness, in the same manner as the iridescent froth conceals the obscure and abysmal archipelago. An emotional compensation of this genre, as I sense it, represents the humoresque of Sancho Panza. In such a prodigious incarnation of serenity and goodness of the soul, Cide Hamete found the balsam and strength, indispensable for pursuing his creative work and discarding somber visions and stinging recollections.

Salve!—then—*Sancho the Pacific, Sancho the Good, Sancho the Jovial!* In the pages of the indeciduous epos, you do not only symbolize the low plateau of common sense, the humble savor of the people that is minted in proverbs, but also the ballast, without which the bloated balloon of the ideal would shatter in the clouds. You are something more and something better than all this. With your grace, cunning and puns, you console the spirit of Cervantes, rendering the overwhelming burden of anguish and misfortune tolerable. Through you he loved life and work, and managed, as time marched on, also cured from his enervating pessimism, to head back to the romantic loves of youth, composing *Persiles*, that true book of chivalry, and *Voyage to Parnassus*, that admirable and definitive literary will. You were soothing balsam to his overexcited sensibility, you saved the genius, and with it, his glory and ours!

More than once, deploring the bitterness distilled from the pages of the Cervantean book, I have exclaimed in a transfer of candid optimism, "Oh! If the unfortunate soldier of Naupactus, fallen and mutilated upon the first encounter, had not had to devour unjust disdains and persecutions; if he had not bemoaned an entire youth wasted in doleful and obscure captivity; if, finally, he had not have to write among the sighs, chortles and blasphemies of the Sevillian underworld, in that infected prison, *where each incommodity reserves its seat...*, how differently, how exhilaratedly and encouragingly would *Don Quixote* have been composed! Perhaps the indeciduous novel would not be the poem of resignation and despair, but the poem of liberty and renewal. And who knows if, in pursuit of the *Caballero of the Lions*, other Quixotes in flesh and blood, influenced by Cervantes' hero, would not have entered into combat as well, in defense of justice and honor, converting, in the end, the joust of lunatics into a glorious campaign of sapients, into a regenerating apostolate, consecrated by the homages of History and the eternal love of Dulcinea..., that ideal woman, whose name—silky and fondling—evokes in the soul the sacred image of the fatherland!..."

But right away, as soon as I fortunately control my unbridled imagination, I become interrupted by a disquieting doubt: "Are you certain," I ask myself, "that *Don Quixote* would have been written, in an environment that was serene and tepid, exempt from grief and miseries?"

And, if it had seen the light in less rigorous conditions of the moral medium, would it be, as it is now, a summary and compendium of human life, and a most faithful historical vision, where, symbolized by universal and eternal types, become agitated, and cry out, all the scourges, poverties and declines of old Spain?

Oh! What a great alarm for the soul, and an instigator of energy, is pain! Similarly to the shoals of sea sparkles *(Noctiluca scintillans)*, whose luminescence increases at the shock of the naval propeller, inert neurons only ignite their light under the whip of painful emotions. Perhaps the privileged cerebrum of Cervantes, in order to attain the tone and ebullience of the sublime inspiration, similarly necessitated the pointed spur of pain and the desolating spectacle of misery.

IT IS TIME now to say a few words about Quixotry. When a literary spirit succeeds in forging a vigorous and universal personality, brimming with life and grandeur, generator of great currents of thought in the realm of society, then the figure of such a fictitious personality burgeons, transcends the boundaries of the fable, invades real life and marks with a special and indelible seal all the people of the race or of the nationality to which the stupendous spiritual creature belongs. That has occurred with the hero of the book of Cervantes.

Many foreigners and not a few Spaniards, thinking that they had discovered a certain air of kinship between the cited protagonist and the moral ambience in which he was conceived, did not restrain from attributing to us, without further inquiries, the disdainful designation of Quixotes, likewise qualifying as Quixotry any Spanish endeavors and aspirations not crowned with success. They take pleasure in portraying us as legendary *Caballeros of the Rueful Countenance*, tenaciously in love with an impossible past and incapable of accommodating to reality and to its useful and salutary teachings.

I would certainly not be the one to refute the complicity that it was nescience, along with the excessive devotion and clinging to the moral and intellectual tradition of our race, that had to do with our sad setbacks and declines; nevetheless, I might be permitted to doubt that ignorance, bewilderment and improvidence constitute the essence and the fundament of Quixotry. Either this word lacks any precise ethical significance or it symbolizes the fervent cult of a lofty ideal of conduct, the will obstinately orientated towards collective light and happiness. Altruistic apostles of peace and social beatitude, true Quixotes feel inflamed by the love of justice, for the triumph of which they sacrifice, without wavering, their very existence, evermore the appetites and delights of sensibility. In all their acts and tendencies they aim not within, in the low regions of some concupiscent soul, but at the spirit of the collective personality, of which they recognize themselves as humble and unselfish cells.

Well now: who, even on the average cognizant of modern History and the habits and trends of the Spanish people of today, would dare qualify us as Quixotes? There were, and there are such among us, no doubt; but, oh! how few, how obscure and disdained!

If I had enough room, I could readily demonstrate how seldom have appeared in our history those spirits that Emerson designates as *representative humans*, and that I would call *humans of the species*, because, clear of low egoisms, to the species they offer themselves and for the species they die. Although it is painful to our soul to confess it, we are forced to acknowledge and to declare that in Spain, excepting its most glorious epochs, the Sanchos abound, while the Quixotes are often lacking.

"How about," one will say, "the Spaniards who discovered and conquered America; those who generously gave their blood fighting for Catholicism in the best part of Europe; those who offered such gallant tokens of confirmed allegiance to their Sovereign and of unblemished devotion to their fatherland, did they not render worship to selflessness nor aspire to an ideal of humanity, of magnanimity, and of justice?"

Indeed, it would be unfair and antipatriotic to disregard that there was a time when Iberia yielded a copious harvest of Quixotes in all sectors of human activity. To this caste would obviously belong several of the discoverers and conquerors of America and Oceania, in whose coarse and ingenuous nature coexisted exquisitely Quixotic characteristics: the devouring thirst for glory, the contempt of life, and the healthy ambition for power and authority; passions which, tempering and ennobling the character, as if emanating directly from Plutarch's *Parallel Lives*, effected true miracles. Among such iron warriors, there were, no doubt, bountiful cruel adventurers, greedy, willing, before anything else, to chase riches and to impose tyranny than to exalt and honor the reputation of the fatherland and her Sovereign. However, two passions, fine avenues to honest Quixotry and even to a somewhat indulgent patriotism, towered above such dissonant and antisocial instincts; namely, the vigor of the untamed will, and the yearning for renown. During those merry times, the capital conquered through heroism was so abundant (although it did not grow afterward, the contrary, it suffered considerable shortfalls), that Spain would stay respected, prosperous and glorious for nearly a century.

Regretfully, those men, in love with life and

action, discoverers and conquerors of immense continents, left a progeny that disparaged the land and was extremely covetous of celestial and beatific islands. Taking refuge in the austerities of religion, fleeing from the world and from its glories, the Quixotes seldom crossed the Atlantic in search of dramatic and romantic feats. The colonies became progressively peopled by Sanchos and, even worse, ruled by Panzas or, at best, by obtuse, mediocre and egocentric *Caballeros of the Green Mantle*. And when the rustic and affable squire ended up being alone, orphaned and nostalgic for the wise counsels and heroic force of Don Quixote, the *Barataria Isles* were gone, and the poor and withered *pegujalero*,[3] reverted to the brownish soiled grounds, remaining bounded, perhaps forever, to the arid Manchean moors...

With all this, are not the art of war and the struggle for geographical expansion, the orders of national activity where the great spurts of the heart and the idealistic spirit are in short supply the most? Undeniably, the domains of art, philosophy and science even further remained orphan of encouraging and exalted

Quixotry.

Despite the opinions somewhat shared by certain critics, historical truth compels the acknowledgment that Spanish art, in its varied manifestations, was essentially human and realistic.[4] As far as poetry is concerned, the national Muse was so balefully disposed to romanticism and hyperbole that, even in the glorious epos of the *romancero* (the assemblage of ballads inspired by the epical deeds of the Reconquest), she would never move past the modest limits of historical narration. As the great authority Menéndez y Pelayo affirms,[5] alluding to the *Poem of the Cid*, "our epic is clean of all chimerical aspiration and is extremely frugal as to the use of the marvelous... The feats attributed to the heroes by the popular Muse are, more or less, the very same that they brought about in the world."

It is evident, on the other hand, that the pastoral poems and the books of chivalry were in their origin exotic productions, tardily inoculated into the national soul and foreign in every aspect to our peculiar literary spirit, which is less remote from classicism than from idealism, knowing to remain loyal, save some romantic and bucolic flirtings, to its intimate, realistic and utilitarian, trend. Only the people, everywhere devoted to the tragic, the marvelous and the implausible, like the eternal child they are, would surrender with ardor to the reading of books and romances of chivalry. The same is happening even today, and will keep happening forever, so long as new social organizations prevent the eternal infant from growing and from reaching maturity with regard to its artistic effectuation.

And yet, what did the very *Don Quixote*, being the work of an incorrigible romantic, represent in its day, leaving aside its intrinsic niceties and sovereign armonies, but the mighty, and essentially conservative, reaction of genuine national realism against extravagated and estranged idealisms?

Further, the present state of Spanish science and philosophy is disorderly, owing to a profuse renunciation and a conceited Quixotry. In love with old books and hardly attentive to the immense spiritual renewal that the Renaissance brought about in every sphere of knowledge, the majority of our thinkers and scientists have confined themselves, as a rule, to modestly implementing the theorems of mathematics and the data in physics and biology discovered by foreigners, in Geography, in Medicine, in the art of navigation, in metallurgy and in the armament industry.[6] With the exception of scholars like Azara, Servet, Gómez Pereira, Huarte, Vives, and some others, in whose minds shone, from time to time, flashes of creative blaze or ingenious intuitions, our scientists always took a stance of disdain for the themes of *basic* research and for the speculative reckonings that were stripped of useful application, being unable to realize, something that occurs among numerous intellectuals at this very moment, that the so-called *applied* science is inseparably connected with the abstract or the idealistic, in the same way as the river is with its source. Odd aberration, spread through routine, and as reprehensible as the florist who surrenders to the mania of plucking the flowers in order to nurture the fruit! How are we to grow the garden of our culture, when we have spent three mortal centuries dismissing or uprooting the flower of the ideas!

An equally deplorable lack of savior Quixotry is observed with sorrow in those domains in which the romantic sentiment and the yearning for fictitious and extraordinary feats are most felicitously associated with the loftiest concerns of civilization and politics. You will discern, no doubt, that I am alluding to the scientific and exploration journeys to which, in better days, the prosperity and renown of the fatherland were due. I wish I were wrong; but I am not aware of any geographic expedition to the North or South Pole launched by Spaniards or Ibero-Americans, while the glorious endeavors of this sort attempted and realized by Yankees, Englishmen, Swedes, Germans, Russians, and even Italians can be counted in the dozens.[7] Sad to confess, but it holds that the pallid *midnight sun* has never highlighted, with its poetic rays, the ripples of the Spanish banner.

At the very gates of the fatherland rises the tenebrous Africa, craddle of the Spanish race as the story goes, according to learned anthropologists. Reclining on the Mediterranean shore, she appears to gaze at us affectionately, like an immense and mysterious sphinx that invites us to eyeball arcane mysteries and contemplate epic endeavors. But in vain is the ingenuous Dulcinea awaiting for centuries the *Caballero of the Lions*. When will the geographer, naturalist or warrior Quixotes arrive at these African beaches, capable of bearing, with the trophies of scientific observation or the tales of romance feats, the single documents of title that cultured peoples today reckon as adequate justification for colonial usufruct?[8]

And turning my attention to more prosaic endeavors: Where are the Quixotes of our industry and commerce? Is it not a painful and distressing spectacle to watch how, at their works, our sumptuous industrialists disdain or dismiss science, the mighty lever, prime mover of the immense manufacturing progress in the present hour, and how they modestly settle (without the slightest inkling as to the long-term foresight that exemplifies prudent egoism) with the sordid import-export of exotic machines and procedures, living, day after day, without struggle and without glory, inside a miserly incubator of tariffs and deals?

Correcting, to the best extent possible, the vices and mental defects of the Spanish race—among which perhaps the most feracious in dire social consequences are the shortage of noble and unselfish civility and the scarcity of sane and exalted Quixotry for the benefit of culture, moral elevation and lasting prosperity of the fatherland—is an act of sublime pedagogy and true regeneration.

Let us admire the book of Cervantes; but let us not divert its morals into domains that the author's spirit did not intend. Realism in art neither refrains from admitting a subtle dose of romantic yeast, in order to excite the interest and elevate the hearts, nor contradicts the supreme and patriotic goal of impressing resolutely idealistic paths on philosophy, science and industry.

Thus, legitimate Quixotry, or in other words Quixotry purified from the filth of ignorance and the outrage of insanity, may find a wide field of applications in Spain: rescuing souls that are fascinated by the dismal cave of error; exploring and exploiting, with a noble national outlook, the boundless riches of the soil and subsoil; deracinating and converting into a delightful and productive garden the impenetrable forest of Nature, where the living agents of disease and death hide threatening; modeling and correcting our own brain with the chisel of intense culture, such that we may render a copious harvest of new ideas and inventions, beneficial for the growth and prosperity of life in all spheres of human activity... There you have the stupendous and glorious adventures reserved for our future Quixotes.

Considered from a moral point of view, nations are supreme compositions of common dreams and aspirations, the sublime blooming of a flora whose multiple rootlets become extended and nourished by all the hearts. I would willingly compare, as well, the great peoples to those poetic coral islands, which emerge from the sea amid the imposing oceanic solitudes. If you contemplate them with the dreamy eyes of the artist, you will be captivated by the merry and placid shores festooned by white froth, by the pilgrim and fragrant flowers, by the colossal trees whose crowns resemble a swaying chorus of celestial birds, thinking that such a paradise surged spontaneously from a strange caprice of Amphitrite; but examine the sub-terrain with the reposed analysis of science, descend to the bed of the sea—which is as priceless as tracking History—and at the marvel of the colossal limestone buttress the work and relics of myriads of infinitesimal and obscure beings: you will then realize that all the grandiose blooming above represents a secular and obstinate construction by innumerable and abnegated existences.

I have spoken.[9]

S. Ramón y Cajal

Tuesday, 9 May 1905

The opening of Cajal's lecture during the festivities marking the tricentennial of the publication of Cervantes' first book of *Don Quijote* in 1605. Left, frontispiece from the chronicle volume published in Madrid [Ramón y Cajal, 1905b]; portrait photo by J. Cao Durán. Right, a later reprint with Cajal's dedication to Marcelino Menéndez y Pelayo (1856–1912), Cantabrian scholar, historian and literary critic, seriously devoted to the history of ideas, Spanish and Latin American literature, and Hispanic philology [Ramón y Cajal, 1905a; Marañon, 1950, p. 11]. As Cajal recollects on his visits to other classes during his excursions through the University [Ramón y Cajal, 1988, p. 452]: "...I shall speak only of Menéndez y Pelayo, who stuttered like Cervantes, but was so learned and eloquent that his students forgot his hesitant speech in gathering the rich honey of his vast and exhaustive literary studies; and of Morayta, Professor of History, who, though he did not attain the richness of diction and of imagination of Menéndez y Pelayo, expounded the history of Greece clearly and methodically, hurling anathemas against the aristocratic, conceited, and disagreeable Lacedaemonians and intoning fervent canticles in praise of democratic Athens, the focus in which ancient culture was concentrated and of which the rays still light European civilization."

NOTES

1 During a celebration of the one hundred years since the award of the Nobel Prize to Cajal and Golgi, Julio Cruz y Hermida, Professor of Gynecology and Obstetrics at the University of Madrid, in an attempt to illuminate the maestro's personality, presented a Cajalian anthology of texts on women from *Charlas de café* as a literary counterpoise to the neuroanatomic publications. Cajal treats the female gender under the prism of beauty, love, marriage, family, even feminism. The anthology concludes with a series of "Insinuations for misogynists" where, in a bright garment of social and philosophical irony, Cajal sways between negative criticism and overt praise, revealing the admiration that his sensitive intellect held for women, reminding "extreme misogynists that even the most ignorant woman from the countryside can give birth to a genius" [Cruz y Hermida, 2006].

2 Literally, *hidalgo* was a Spanish nobleman by birth, originally a land owner; *caballero* was a Spanish knight by social status, or one who had made use of the military privileges of the class [de Cervantes Saavedra, 1979, p. 207, fn. 6]

3 [Rural worker, belonging to a marginal group of the Cartagena and Lorca fields during the *Ancien Régime*, who would lease a small piece of land for farming on a short-term contract.]

4 Remember that our glorious painters—Velázquez, Zurbarán, Coello, Rivera, Goya, etc.—were, before anything else, great observers of Nature. Even in the very Murillo, the most mystical of all, realism surpassed idealism.

5 Salillas also cites these important views of Menéndez y Pelayo, justifying the thesis whereby the national soul, heroic, sane and robust during our golden age, degenerated somewhat later into the boast and impotence of thuggery and naughtiness.

6 It is just and patriotic to proclaim that science in Spain during the 16th century established many causes and glimpsed at bright and productive truths. But, unfortunately, very few theories were completed or perfected, because her scholars lacked, with the longing for international glory, an eminently Quixotic passion, a supraintensive attentive care, and an indefatigable perseverance. It is sad to realize that philosophers, as enlightened as Gómez Pereira, Vives, Francisco Vallés, Fox Morcillo, etc., who, formulated, before anyone else, the principles of the experimental method, but without demonstrating its usefulness in the realm of facts; that the famous Arias-Montano explained the rise of water in a pipette by the atmospheric pressure, without arriving, however, at the laws of Torricelli and Pascal; that Pérez de Oliva, professor of optics and magnetism in Salamanca (1533), mentioned the possibility of applying magnetism for the communication

among absent and distant persons, but without, no doubt, any important discovery in the field; that Pedro de Liria guessed the existence of a magnetic pole at a distance of a few degrees from the geographic pole, without specifying, by means of sufficient observations, its position; that Juan de Escribano, the translator of the *Porta*, was content with foreshadowing the practical importance of the elastic force of steam, etc., etc. What contributed to such a modesty of theoretical fruits was, no doubt, the encyclopedic mania, which, when raised above reason, creates peaks to dominate wide horizons, also dwarfing the objects, clothing them with mist. Encyclopedists, on a par with the great pedagogues and commentators, were the cited Arias-Montano, El Broncense, Pedro Ciruelo, Nebrija, the astronomer Santa Cruz, etc., and precisely, because of that, they did not give rise to discoveries worthy of their genius.

Not for lack of observation or patience, but because of excess utilitarianism, pilots like Juan de la Cosa, geodesists like Esquivel, metallurgists like Medina, Vargas, and Alonso Barba, zoologists like Acosta and Oviedo, and botanists like Hernández, among others, fell short of founding modern cosmography, chemistry, zoology and botany.

7 Although such endeavors, wasteful at first sight, did not lead to the solution of any interesting geographical, meteorological or physical problem, they always constituted an exercise in heroism, indispensable for

weak nations in avoiding relegation to the villainy of rude utilitarianism, and in commanding respect for the Quixotes of military glory.

8 Also, with regard to the exploration of Africa, we have forgotten glorious traditions. Besides the famous journey that the Catalan Domènec Badia i Leblich ("Alí Bey al-Abbassí") realized, early last century, through Barbary, Tunisia and Arabia, mention should be made to the exploration of North Africa that was undertaken in the 16th century by the soldier and geographer Luis del Mármol y Carvajal. It seems that, nowadays, the Spanish Society of Natural History intends to remedy the serious error which Spain committed by abandoning, during the entire 19th century, the scientific exploration of Africa to the Quixotry of the English, the French and the Germans.

9 *Εἴρηται τὰ παρ' ἐμοῦ, καὶ παύομαι.* "He dicho" [Ramón y Cajal, 2004] or "I have spoken what I have to say, and I rest my case." [The clause is traced to the *Apology of Palamedes* by the Presocratic philosopher Gorgias of Leontinoi (c. 483–375 BC), verse 37.]

BIBLIOGRAPHY

Cannon, D.F. (1949) *Explorer of the Human Brain: The Life of Santiago Ramón y Cajal (1852–1934)*. Henry Schuman, New York.

Cannon, D.F. (1951) *Vida de Santiago Ramón y Cajal: Explorador del cerebro humano* (traducción de A. Folch y Pi). Biografías Gandesa – Exportadora de Publicaciones Mexicanas, México.

Clarke, E., Jacyna, L.S. (1987) *Nineteenth-Century Origins of Neuroscientific Concepts*. University of California Press, Berkeley.

Courville, C.B. (1953) Santiago Ramón y Cajal (1852–1934). In: Haymaker, W., Baer, K.A. (eds.) *The Founders of Neurology*. Charles C. Thomas Publisher, Springfield, IL, p. 74–7.

Craigie, E.H., Gibson, W.C. (1968) *The World of Ramón y Cajal, with Selections from His Nonscientific Writings*. Charles C. Thomas Publisher, Springfield, IL.

Cruz y Hermida, J. (2006) La figura de la mujer en el pensamiento y la obra de Cajal. *Anales de la Real Academia Nacional de Medicina 123:* 689–710.

de Cervantes Saavedra, M. (1979) *Don Quixote—The Ingenious Gentleman of La Mancha* (translated by J. Ormsby). Easton Press, Norwalk, CT.

DeFelipe, J. (2005) Cajal y sus dibujos: Ciencia y arte. In: Martín Araguz, A. (ed.) *Arte y neurología*. Editorial Saned, Madrid, p. 213–30.

Durán Muñoz, G., Sánchez Duarte, J. (1945) *Recopilaciones y estudios Cajalianos: La psicología de los artistas, las estatuas en vida y otros ensayos inéditos o desconocidos de*

Santiago Ramón y Cajal. Industrias Gráficas Ortega, Vitoria-Gasteiz.

Escribano, A. (1924) Santiago Ramón y Cajal. *Revista de Segunda Enseñanza — 2.ª Época (Madrid) 2:* 234–5.

Hoare, M.R. (1998) Ramón y Cajal's testament to old age. *Reviews in Clinical Gerontology 8:* 163–71.

Ibarz Serrat, V. (1994) *La psicología en la obra de Santiago Ramón y Cajal*. Institución Fernando el Católico, Zaragoza.

Jakob, C. (1935) Santiago Ramón y Cajal: La significación de su obra científica para la Neuropsiquiatría. *La Semana Médica (Buenos Aires) 42:* 529–536.

Jakob, C. (1938) El significado de la obra de Ramón y Cajal en la filosofía de lo orgánico. *Humanidades (La Plata) 26:* 237–55.

López Piñero, J.M., Terrada Ferrandis, M.L., Rodríguez Quiroga, A. (2000) *Bibliografía Cajaliana: Ediciones de los escritos de Santiago Ramón y Cajal y estudios sobre su vida y obra*. Albatros – Artes Gráficas Soler, Valencia.

Marañon, G. (1950) *Cajal, su tiempo y el nuestro*. Antonio Zuñiga Editor, Santander – Madrid.

McMenemey, W.H. (1969) The World of Ramón y Cajal —With Selections from His Nonscientific Writings, by E. Horne Craigie and William C. Gibson (book review). *Journal of Neurology Neurosurgery and Psychiatry 32:* 254.

Palma, J.-A., Palma, F. (2012) Neurology and Don Quixote. *European Neurology 68:* 247–57.

Pasqualini, C.D. (1999) Cien años después en investigación científica. *Medicina (Buenos Aires) 59:* 798–800.

Ramón y Cajal, S. (1905a) *Psicología de don Quijote y el quijotismo* (Discurso leído en la sesión conmemorativa de la publicación del Quijote, celebrada por el Colegio

Médico de San Carlos el día 9 de Mayo, 1905). Nicolás Moya, Madrid.

Ramón y Cajal, S. (1905b) Psicología del Quijote y el quijotismo. In: Sawa, M., Becerra, P. (eds.) *Crónica del centenario del don Quijote*. Establecimiento Tipográfico de Antonio Marzo, Madrid, 1905, p. 161–8.

Ramón y Cajal, S. (1905c) Sobre la psicología de don Quijote de La Mancha y el quijotismo. *Boletín del Colegio de Médicos de Gerona 10:* 101–13.

Ramón y Cajal, S. (1906) *Relación de los títulos, méritos y trabajos científicos con el retrato del autor* (Publicada á expensas de la Facultad de Medicina). Imprenta y Librería de Nicolás Moya, Madrid.

Ramón y Cajal, S. (1944) *La mujer — Psicología del Quijote y el quijotismo*, 3rd edn. J. García Perona Editor – Gráficas Sebastián, Madrid.

Ramón y Cajal, S. (1954) *La psicología de los artistas* (Prólogo, compilación y notas de G. Durán Muñoz y J. Sánchez Duarte—Colección Austral*, vol. 1200). Espasa-Calpe Argentina, Buenos Aires.

Ramón y Cajal, S. (1961) *Obras literarias completas*, 4th edn. (Con una nota preliminar de F.C. Sáinz de Robles). Aguilar, S.A. de Ediciones – Vicente Mas, Madrid.

Ramón y Cajal, S. (1988) *Recollections of My Life* (translated by E.H. Craigie and J. Cano). The Classics of Neurology and Neurosurgery Library – Gryphon Editions, Birmingham, AL.

Ramón y Cajal, S. (2004) La psicología de Don Quijote de La Mancha y el Quijotismo. *Arbor (Madrid) 179:* 1–12.

Ramón y Cajal, S. (2012) *Psicologia del don Quijote e il quijotismo* (translated and annotated by P. Piro). Mimesis

Edizioni (Il Caffè dei Filosofi, no. 35), Sesto San Giovanni, Milano.

Ramón y Cajal Junquera, S. (2000) *Santiago Ramón y Cajal*. Caja de Ahorros de la Inmaculada, Aragón.

Ramón y Cajal Junquera, M.Á. (2002) Cajal en Barcelona: Santiago Ramón y Cajal y la hipnosis como anestesia. *Revista Española de Patología 35:* 413–4.

Romero, A. (1984) *Fotografía Aragonesa/1: Ramón y Cajal*. Diputación Provincial de Zaragoza, Zaragoza, p. 82–5.

Sherrington, C.S. (1935) Santiago Ramón y Cajal 1852–1934. *Obituary Notices of Fellows of the Royal Society 1:* 424–41.

Tabarés-Seisdedos, R., Corral-Márquez, R. (2001) Miguel de Cervantes, 1547–1616. *American Journal of Psychiatry 158:* 9.

Taylor, D.W. (2008) Ramón y Cajal, Santiago. *Complete Dictionary of Scientific Biography.* 2008. Accessed from *Encyclopedia.com* on 15 February 2014.

Triarhou, L.C. (2015) *Cajal Beyond the Brain: Don Santiago Contemplates the Mind and Its Education.* Corpus Callosum, Thessalonica - Indianapolis.

Williams, H. (1954) *Don Quixote of the Microscope: An Interpretation of the Spanish Savant Santiago Ramón y Cajal (1852–1934).* Jonathan Cape, London.

Williams, H. (1955) *Don Quijote del microscopio: Una interpretación del sabio español Santiago Ramón y Cajal (1852–1934)* (traducción de P. Abelló). Taurus, Madrid.

www.ingramcontent.com/pod-product-compliance
Lightning Source LLC
Chambersburg PA
CBHW020711180526
45163CB00008B/3029